climbing feet
koala

digging feet
squirrel

quiet feet
bunny

skinny feet
spider

hard feet
cow

fat feet
lion

four wet feet
polar bear

For my parents, with love and thanks—ST

Library of Congress Cataloging-in-Publication data is on file with the publisher.

Text copyright © 2019 by Sue Tarsky
Illustrations copyright © 2019 by Albert Whitman & Company
First published in the United States of America in 2019 by Albert Whitman & Company
ISBN 978-0-8075-9039-3

Printed in China
10 9 8 7 6 5 4 3 2 1 HH 22 21 20 19 18

Design by Aphee Messer

For more information about Albert Whitman & Company,
visit our website at www.albertwhitman.com.

100 Years of Albert Whitman & Company
Celebrate with us in 2019!

Whose Are These?

Whose Feet?

Sue Tarsky

Albert Whitman & Company
Chicago, Illinois

small feet

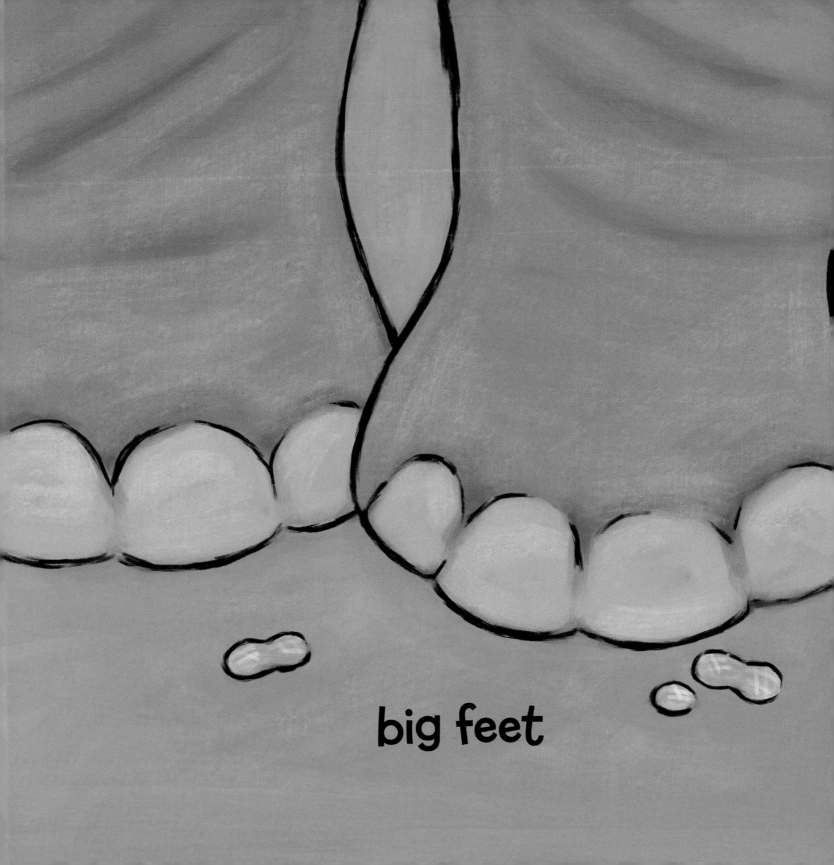

big feet

fat feet

skinny feet

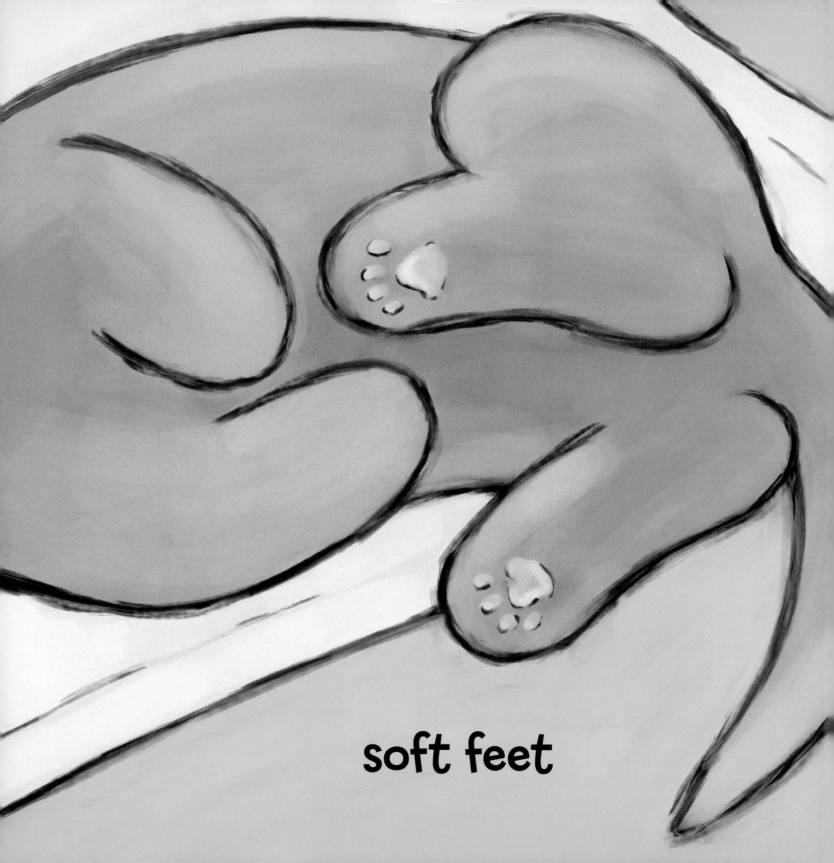

soft feet

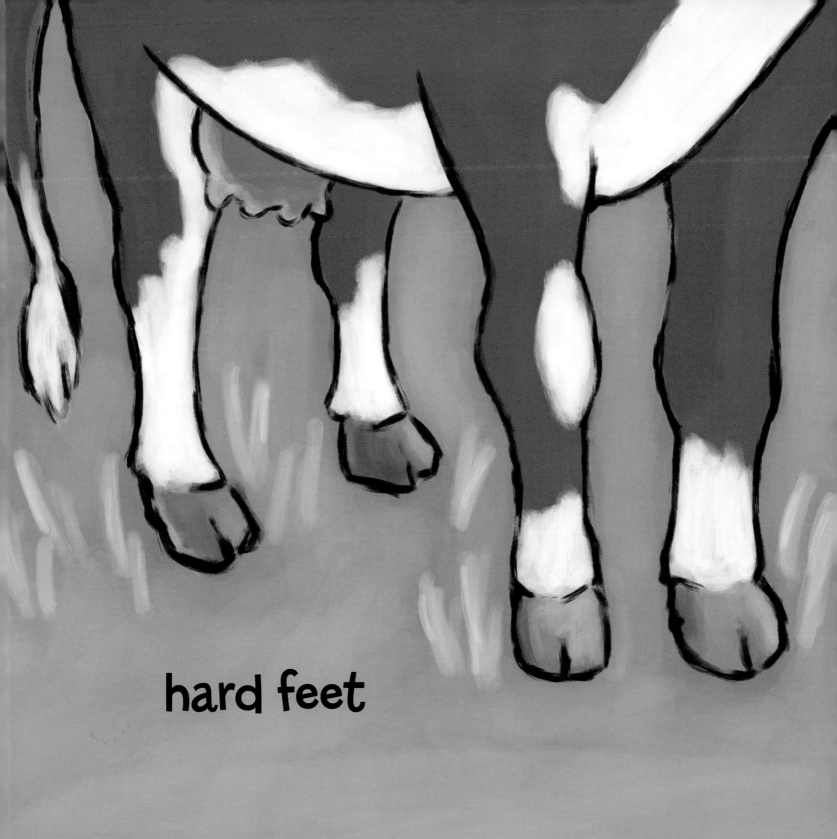

hard feet

digging feet

climbing feet

quiet feet

noisy feet

slow feet

hurry-up feet

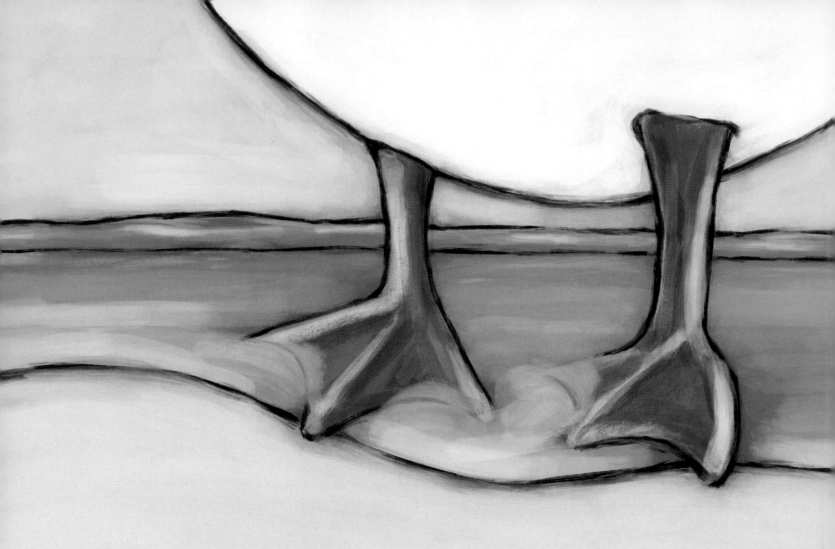

two wet feet

four wet feet

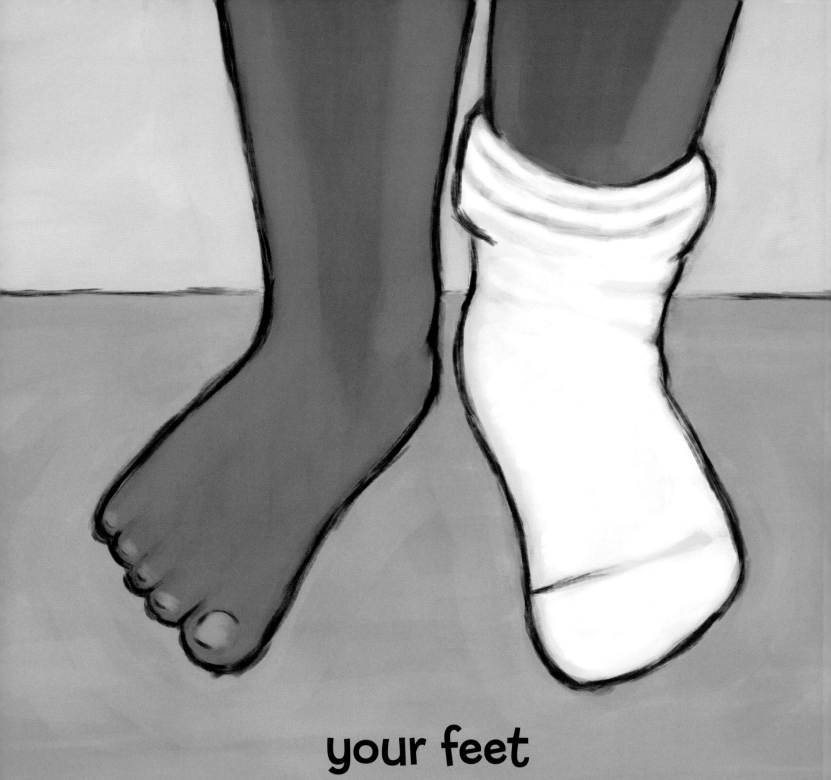

your feet

my feet!

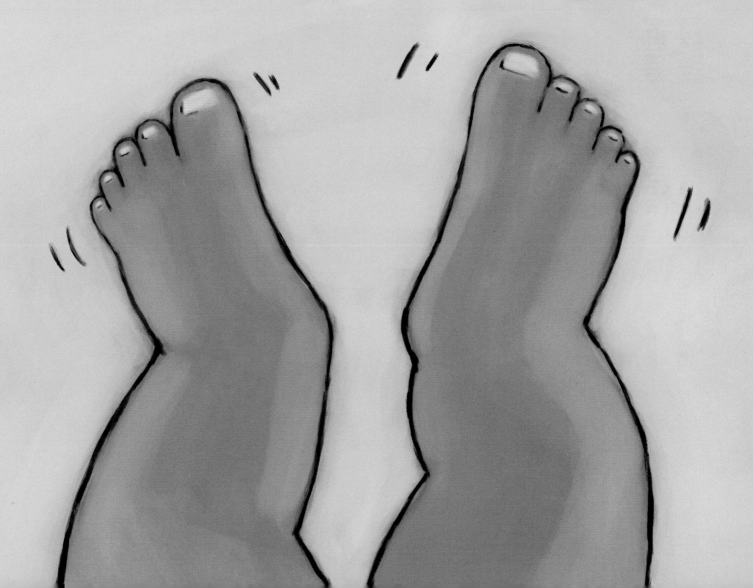

two wet feet
duck

big feet
elephant

noisy feet
horse

small feet
mouse

soft feet
kitten

slow feet
tortoise

hurry-up feet
dog